THIS BOOK BELONGS TO

_____

DEDICATION

This book is dedicated to my daughter, Maia, for teaching me that being stubborn just means you are passionate about something you love.

"THE ONLY PERSON WHO CAN STOP YOU FROM REACHING YOUR GOALS IS YOU."
JACKIE JOYNER-KERSEE

IN THIS BOOK WE WILL LEARN THE AMAZING STORIES OF

Drs. Rita Levi-Montalcini and Eve Marder

THESE AMAZING WOMEN ASKED QUESTIONS AND WERE CURIOUS. THEY NEVER LET ANYONE STOP THEM.

PART ONE:
THE NERVE GROWTH PIONEER

THE REMARKABLE JOURNEY
OF DR. RITA LEVI-MONTALCINI

A CURIOUS GIRL

Once upon a time in Turin, Italy there lived a curious girl named Rita. She loved asking questions and exploring the world around her. Her curiosity would take her on an incredible journey of discovery.

A CURIOUS YOUNG WOMAN

Rita went to college to study medicine and become a doctor. But her true passion lay in understanding the brain, our body's most mysterious organ.

FLEEING FROM DANGER

A map of Italy showing Rita's journey from Turin to Florence

During World War II, Rita faced danger because of her Jewish ancestry. She had to hide in Florence, Italy to stay safe. But even in hiding, her mind never stopped working.

THE HIDDEN SCIENTIST

She created a lab in her bedroom and dreamed of understanding how nerves grow and connect.

INCREDIBLE DISCOVERY

After the war, Rita continued her research. Rita discovered something incredible: there is a special chemical in a mouse tumor which made nerves grow. She called this chemical, Nerve Growth Factor (NGF).

TINY MESSENGERS

Imagine tiny messengers inside your brain, telling nerve cells how to grow and connect. NGF was like a magical potion that helped nerves grow and form networks.

THE NOBEL PRIZE

In 1986, Rita received the Nobel Prize in Physiology or Medicine for her ground-breaking discovery.

She was 77 years old when she received this award.

THE NOBEL LAUREATE

A person who wins the Nobel Prize is known as a Nobel Laureate. She was the first Nobel Laureate in history to reach 100 years of age.

A ROSE IS BUT A ROSE

Rita also earned one of Italy's highest honors when she became a senator for life.

Her most unique honor may be having a rose named after her. The "Rita Levi-Montalcini" grows in a cluster of fragrant apricot flowers with creamy-pink backgrounds.

INSPIRING OTHERS

Rita's work inspired other scientists to explore the brain too. Her finding helped scientists learn more about brain diseases like Alzheimer's and Parkinson's, hoping to find cures.

A LEGACY OF HELPING OTHERS

Rita's legacy lives on. NGF is now used to treat nerve injuries and brain disorders. She taught us that curiosity and persistence can change the world.

RITA SHOWED THE WORLD
THAT OUR BRAINS ARE LIKE
INTRICATE FORESTS, WITH
NERVES REACHING OUT LIKE
BRANCHES.

DR. RITA LEVI-MONTALCINI'S STORY TEACHES US TO NEVER GIVE UP.

NO MATTER HOW HARD IT MAY GET

PART TWO: LESSONS FROM THE LOBSTER

DR. EVE MARDER'S GUT FEELING

LOBSTER LOVE

Eve loved lobsters. Not because she wanted to eat them, but because their tiny stomachs held a big mystery.

SPECIAL BRAIN

Eve put on her lab coat and peered inside the
lobster. She saw the lobster has a special brain that
allowed it to sense its' environment.

ENDLESS DETERMINATION

Eve studied the lobster brain day and night. She discovered that these 30 neurons weren't just simple switches.

THE LOBSTER DANCE

They danced with each other, passing messages like secret notes. And guess what? They could change their tune!

THE LOBSTER SYMPHONY

Eve listened carefully. The neurons hummed like a tiny orchestra. Some played fast notes, telling the lobster to wiggle its tail.

NEURAL TUNES

Others played slow tunes, saying time for lunch!

This made Eve wonder how these neurons work

together.

CRACKING THE CODE

One day, Eve cracked the code. The lobster neurons were like friends at a party: they chatted, laughed, and swapped dance moves.

SPECIAL HELPERS

Do you know what made them change? Special helpers called "neuromodulators." These tiny molecules turned the lobster brain into a flexible dance floor!

THE BRAIN PUZZLE

Eve's lobster lessons taught her about all brains, big and small. And just like the lobster neurons, our brains could change too!

STELLAR BEAUTY

She realized that our human brains were like galaxies, full of stars (neurons) and constellations (connections).

EVE SHOWED EVEN A LOBSTER BRAIN CAN HOLD COSMIC WONDERS

DR. EVE MARDER'S STORY TEACHES US THAT EVEN SMALL DISCOVERIES CAN HAVE A BIG IMPACT!

AND IT ALL STARTED WITH A QUESTION.

NEVER LOSE YOUR WONDER. MAYBE YOU WILL MAKE THE NEXT AMAZING DISCOVERY.

YOUR BRAIN IS A UNIVERSE WAITING TO BE DISCOVERED

ABOUT the Author

Seena Mathew loves encouraging children to pursue their love of science and the brain. She has always been fascinated by the brain. Have you ever wondered why some people can remember everything they read? Or wondered why some people have difficulty seeing certain colors? She did too and spent her years in college learning all about it. Seena is passionate about educating people on the complexities of the brain and hoping they are as interested in it as she is.

All information from this book was obtained from peer reviewed journals and sources available on the internet.

All images were created using artificial intelligence.

These are unofficial biographies of the individuals mentioned.

www.ingramcontent.com/pod-product-compliance
Lightning Source LLC
Chambersburg PA
CBHW042039230526
45474CB00005B/22